LOW-CARBON ENERGY PATHWAY FOR DENPASAR CITY, BALI, INDONESIA

DECEMBER 2024

ASIAN DEVELOPMENT BANK

ADB

6 ADB Avenue, Mandaluyong City, 1550 Metro Manila, Philippines
Tel +63 2 8632 4444; Fax +63 2 8636 2444
www.adb.org

ISBN 978-92-9277-158-4 (print); 978-92-9277-159-1 (PDF); 978-92-9277-160-7 (ebook)
Publication Stock No. TCS240626-2
DOI: http://dx.doi.org/10.22617/TCS240626-2

Note: In this report, "$" refers to Unoted States dollars.

Cover design by Mike Cortes.

Contents

Tables and Figures

Tables

Figures

Acknowledgments

The *Low-Carbon Energy Pathway for Denpasar City, Bali, Indonesia* is an in-depth analysis commissioned by the Asian Development Bank (ADB) and developed by the Energy Sector Office, Sectors Group. The study uses the MESSAGEix model to create a low-carbon pathway for the city, identifying key areas for development and technology to reduce energy demand and emissions. It also estimates the investment needed over 30 years (2020–2050).

The study was carried out under the leadership of Kee-Yung Nam, principal energy economist, with supervision from Priyantha Wijayatunga, senior director of the Energy Sector Office, Sectors Group. The report was prepared and the modelling was conducted by the management and staff of PricewaterhouseCoopers International Limited. David Morgado, senior energy specialist, coordinated and reviewed the content. John Weiss served as the technical editor.

A team of both international and national experts offered significant insights as peer reviewers. Maria Dona Aliboso, associate project officer, and Maria Carmela Abadeza, associate project analyst, managed the production process and provided analytical and administrative support. Ma. Theresa Mercado copyedited the report, while Lumina Datamatics did the layout. Jess Macasaet conducted the proofreading and page proof checking of this publication. The study also gained valuable insights and feedback from colleagues at the ADB Energy Sector Office, Sectors Group.

This study is supported by funding from the Clean Energy Fund under the Clean Energy Financing Partnership Facility, the People's Republic of China Poverty Reduction and Regional Cooperation Fund, the Urban Climate Change Resilience Fund, and the E-Asia Knowledge Partnership Fund.

Abbreviations

BAU	business-as-usual
GHG	greenhouse gas
HVAC	heating, ventilation, and air-conditioning
LPG	liquefied petroleum gas
PV	photovoltaic
TFED	total final energy demand
TPES	total primary energy supply

Weights and Measures

GW	gigawatt
$MtCO_2$	million tons of carbon dioxide
$MtCO_2e$	million tons of carbon dioxide equivalent
Mtoe	million tons of oil equivalent
MW	megawatt
TWh	terawatt-hours

Executive Summary

This study sets out a low-carbon pathway for Denpasar, the capital city of the island of Bali, Indonesia. In recent decades, Denpasar has developed into a major tourist hub with hotels, resorts, restaurants, and other allied businesses set up to support a growing tourism demand and contributing to the rising energy demand of the city. Currently, most of the city's energy is imported from outside the city boundaries, as oil or electricity. The majority of electricity demand is met from the Java-Bali electricity grid, although lately a small amount of off-grid solar photovoltaic (PV) supply has emerged from within the city.

The study is based on a detailed bottom-up analysis using the MESSAGEix model to develop a low-carbon pathway for the city. The objective is to identify the key areas for low-carbon development and technology intervention and their impact on the energy demand and overall emissions of the city's energy system. Based on the energy system analysis approach, the study estimates the investment required by this pathway over the course of 30 years (2020–2050).

The study uses a scenario-based approach applying two alternatives: business-as-usual (BAU) scenario and low-carbon scenario. The former assumes continuation of current and historic policy and technology trends, while the latter incorporates policy and technology changes to support the Paris Agreement climate goals. The total final energy demand (TFED) is projected to grow from 0.31 Mtoe in 2020 to approximately 2.02 million tons of oil equivalent (Mtoe) in 2050 in the BAU scenario and to 1.81 Mtoe in 2050, in the low-carbon scenario. Hence TFED is approximately 10% lower under the low-carbon scenario compared to the BAU. This is because efficient utilization of energy is an integral element of decarbonization. With a rise in uptake of energy-efficient solutions, final energy demand from all sectors is projected to fall would over time relative to the BAU trajectory, resulting in energy and emissions savings. Of the reduction in TFED, relative to the BAU scenario, a majority (73%) of the reduction takes place in the transport sector, due to the spread of electrification of vehicles, followed by 23% from the residential non-cooking sector, with more efficient building design and cooling, and 4% from residential cooking, with greater use of electric devices.

The reduction in final energy demand will directly influence the primary energy supply mix of the city. The total primary energy supply (TPES) is estimated to grow from 0.35 Mtoe in 2020 to 2.3 Mtoe in 2050 under the BAU scenario and to 2.06 Mtoe by 2050, in the low-carbon scenario. In addition to the reduction in energy demand, there is also fuel switching in the energy mix. A very slow growth in solar photovoltaic (PV)-based electricity is envisaged under the BAU scenario. In the low-carbon scenario, growth in solar PV-based capacity addition is estimated to be higher, contributing 3% to generation in 2050. The share of the remaining fuels in the TPES does not differ markedly between the two scenarios and the city area will remain dependent on imports, either of electricity from the national grid or oil, in both.

With economic growth an increase in overall greenhouse gas (GHG) emissions is projected under both scenarios. Under the BAU scenario, GHG emissions associated with energy use within the city are estimated to reach 4.8 million tons of carbon dioxide ($MtCO_2$) in 2050, while in the low-carbon scenario they are 0.85 $MtCO_2$ lower, at 3.95 $MtCO_2$ in 2050. The 18% reduction in emissions (relative to the BAU) is due primarily to energy-efficient interventions on the demand side and, to a limited extent, to increased penetration of cleaner energy in the overall energy supply mix.

The low-carbon scenario highlights the potential for energy efficiency interventions in the city. While demand for space cooling is expected to grow with economic growth, the increased use of efficient air-conditioning systems, like inverter air conditioners and district cooling, combined with green building design, can limit the growth of energy demand and emissions in buildings. Extensive electrification of end uses is also possible in transport through electric vehicles, and residential cooking through electric cooking.

There are multiple benefits associated with electrification, including efficient utilization of energy, no emissions at the point of usage, and the possibility of using decentralized and local renewable energy generation. However, they come at a cost, as they require additional investment. Under the BAU scenario, simply to allow the expansion of the energy system on its historical trend the model estimates a total cost of investment of $3.06 billion (2010 US dollars) over 2020–2050. With additional interventions under the low-carbon pathway, the total cost is 16% higher at $3.54 (2010 US dollars), a difference of nearly $0.5 billion over the same period.

Hence, simply expanding the system on its current trajectory will require substantial funding. Moving to a low-carbon pathway poses further challenges. The implementation of low-carbon alternatives, as envisaged under the low-carbon scenario, will be guided by national policies, but in addition, these must be supported by the formulation of appropriate policies and regulations at the city level. Local initiatives will be required to establish partnerships between investors, the city administration, and the regional and federal governments. Critically, there is a need to influence consumer behavior leading to the adoption of cleaner and more energy-efficient alternatives, whenever these are available. This requires an effective communication strategy, setting out the benefits and mechanisms for incentivizing cleaner products and energy-efficient appliances and dis-incentivizing the emission-intensive alternatives.

1. Introduction

Denpasar City in Bali, Indonesia, has been selected as the case study of a low-carbon pathway for an urban area. Denpasar is Bali's capital and most populous city. Over the past few decades, it has developed into a tourist hub, with infrastructure and other allied businesses set up to support a growing tourism demand. In terms of climate policy, Denpasar has recently passed a regulation (Governor Regulation Number 45) that directs the city to meet energy demand with clean energy including solar, wind, geothermal, and biomass. The city is signatory to the Global Covenant of Mayors and has an integrated Climate Action Plan to contribute to national efforts toward decarbonization. The major challenge that the local government faces is tapping into the abundant solar resources of the area to transition Denpasar to cleaner fuels and reduce dependence on energy imports from the national grid.

Methodology

The model used for this study is the MESSAGEix modeling framework developed by the International Institute for Applied Systems Analysis. MESSAGEix is a versatile, open-source, dynamic systems-optimization model providing a framework for analyzing an energy system with its interdependencies, allowing for resource endowments and resource potential, extraction rates, imports, exports, the generation of electricity, and the conversion of fuels to energy for end use. The optimization model gives the least-cost solution subject to constraints over predefined time periods.

The MESSAGEix model is widely used for assessments of national and subnational energy transitions. It has the flexibility to evaluate and assess mitigation potential across different sectors for various low-carbon development scenarios and allows the impact of changes in policies and technologies to be assessed. The sectors considered for this study are commercial, industrial, residential, and transport. The sector end-use energy demands are supplied exogenously to the model. The outcomes of energy demand are presented at the final energy level, which is one level before the end use level in a typical Reference Energy System. The process of energy flow from final energy level to end use level takes into consideration the inefficiencies of fuel use in sectors.

To ascertain the energy transition outlook, two scenarios are defined:

- BAU scenario–assumes continuation of past and current trends in the future. This scenario derives future patterns in technological penetration and resource consumption from historical trends and known policy shifts.
- Low-carbon scenario–assesses options for achieving energy and emissions reduction trajectories based on the assumption of an enhanced ambition for low-carbon development to support the Paris Agreement goals agreed by the government.

These scenarios are not purely cost-optimization scenarios as a certain level of technological guidance has been adopted to maintain consistency with current and historic trends in the case of the BAU scenario and with the enhanced low-carbon policy initiatives in the case of the low-carbon scenario.

The national technology road map of Indonesia, developed under the Asian Development Bank technical assistance 9690 is taken as the starting point.[1] The approach requires creating a city-level energy database from data in the public domain and from consultation with local experts and consultants and then building a city-level model. The energy demand and supply database were developed and extended through a series of virtual consultation seminars conducted with local parties. The approach is illustrated in Figure 1.

Figure 1 Schematic Illustration of the Overall Approach

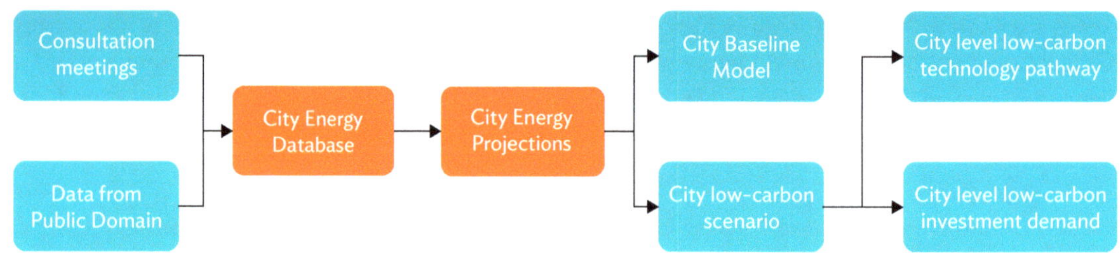

Source: Authors.

In a few areas, the data available at the city level were limited and this necessitated approximations. For example, heating, ventilation, and air-conditioning (HVAC) data are only considered for residential buildings due to the unavailability of local information on cooling demand required for commercial buildings. Furthermore, fuel oil consumption has been incorporated under residential cooking as no data are available on its use by the commercial sector. Similarly, different uses of electricity are aggregated under the residential and commercial sectors.

Figure 2 has a schematic representation of the city modeling highlighting the key energy users.

In terms of the scope of the model, the broad sector assumptions used are summarized in Figure 3.

Business-as-Usual Scenario

The BAU scenario draws on a combination of factors to project a future energy and emissions trajectory, based on the objective of cost minimization, historic trends in installed capacities and utilization, resource potential, and growth in end-use energy demand from the commercial, industrial, residential, and transport sectors. The scenario projections for energy use are based primarily on the current and historic utilization patterns of the energy commodities—coal, oil, gas, biomass, hydro. The uptake of emerging renewable and cleaner technologies such as solar, wind, and nuclear is determined by their cost without any constraint imposed by a policy shift.

1 Integrated High Impact Innovation in Sustainable Energy Technology - Energy System Analysis, Technology Road Maps and Feasibility Studies for Pilot Testing (Subproject 1). https://www.adb.org/projects/52041-002/main.

Figure 2 Schematic Illustration of City Modeling by Key Users

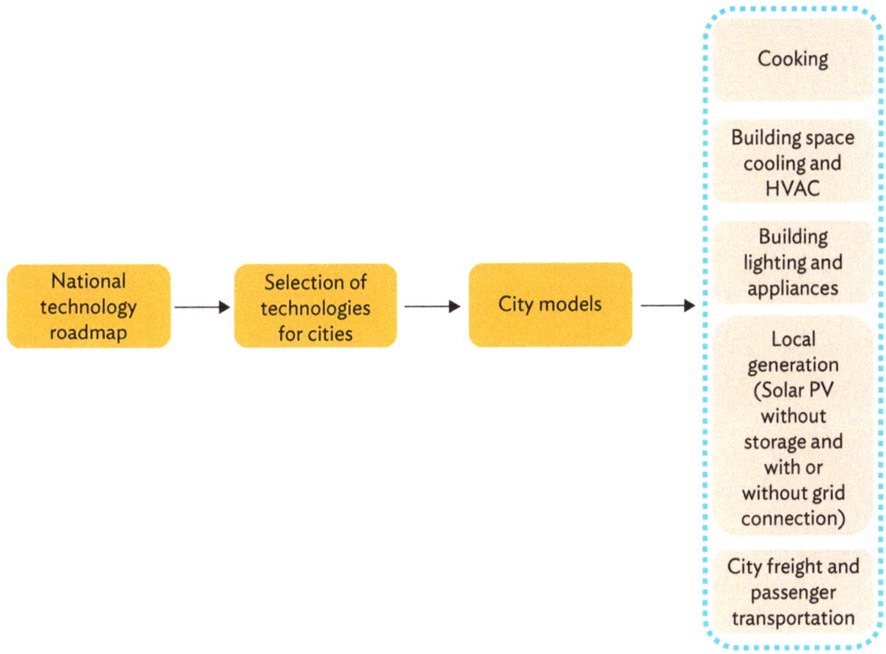

HVAC = heating, ventilation, and air-conditioning; PV = photovoltaic.

Source: Authors.

Figure 3 Key Assumptions

- Primary energy supply is taken as import of commodities through national supply chain
- Agriculture sector is out of scope as it comes under the national jurisdiction
- Majorly road-based passenger and freight transportation is considered due to data limitation. Inland water transport at very small scale
- Population-based indexing of passenger kilometers covered for categories of road passenger vehicles like cars, two-wheelers etc. to estimate the end-use demand of vehicle utilization
- All electricity is assumed to be imported through national grid, excluding potential renewable installations like solar PV

PV = photovoltaic.

Source: Authors.

Low-Carbon Scenario

The low-carbon scenario takes into consideration a set of policy and technology measures on both the supply and demand sides to steer the economy toward a low-carbon pathway. These measures include demand-side electrification, supply and demand-side efficiency improvements, and fuel-switching to carbon neutral or low-carbon fuels such as natural gas, biofuels, and off-grid renewable energy generation. The nationally determined contribution climate targets are not explicitly provided as an input, but rather the scenario outcomes are used to assess the potential of such interventions to shift the economy toward these targets.

Figure 4 summarizes the approach used to estimate energy demand from each sector, with assumptions about policies varying between the two scenarios.

Figure 4 Approach for Estimating Energy Demand

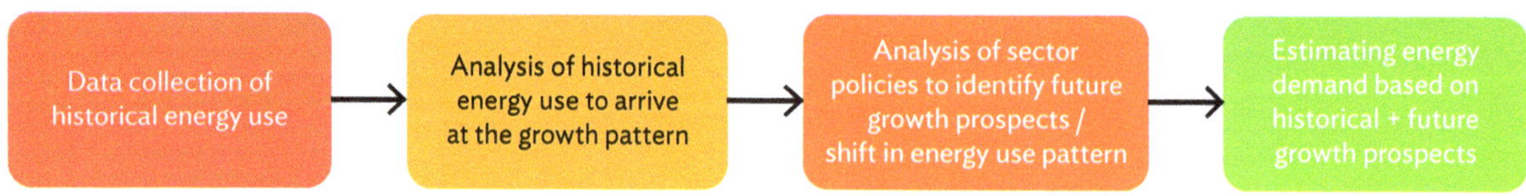

Source: Authors.

Demand-side management or behavioral interventions assumed in the low-carbon scenario, such as fuel switching, are reflected at the useful energy or end-use level. The final to end-use energy conversion efficiencies used reflects the process and/or technological efficiencies built into the analysis. The list of low-carbon targets adopted for the low-carbon scenario is in Table 1.

Table 1 Low-Carbon Targets in the Low-Carbon Scenario

Scenario	Target Description
Fuel shift	More than 25% of the residential cooking would be met through electric cooking technology (e.g., induction cookers) at the final energy level by 2050
	More than 15% electric vehicle penetration in the road transport sector by 2050 at final energy level
	Biofuel blending of more than 20% in the conventional petroleum supply for transport by 2050
Renewable and new energy	About 10% of total electricity demand in residential buildings (excluding HVAC demand) met by solar rooftop PV in 2050
	7% of the total electricity demand in commercial buildings (excluding HVAC) met by solar rooftop PV by 2050
Energy efficiency	More than 80% of total cooling demand will be met through centrifugal chillers by 2050

HVAC = heating, ventilation, and air-conditioning; PV = photovoltaic.

Source: Compiled by authors.

The model requires projections for population and income growth for the city that are common to both scenarios. Table 2 and Table 3 provide an overview of the projections used. They imply a gross domestic product per capita for the city in 2020 of $4,452 (in 2020 prices) projected to increase to $21,092 by 2050.

Table 2 Population Projections for Denpasar (millions)

2020	2025	2030	2035	2040	2045	2050
0.96	1.08	1.11	1.14	1.17	1.2	1.23

Source: Compiled by authors.

Table 3 Gross Domestic Product Projection for Denpasar (Rp trillion 2010/yr)

2020	2025	2030	2035	2040	2045	2050
32.75	33.07	47.34	67.78	97.02	138.9	198.84

Source: Compiled by authors.

Policy Framework

The Government of Indonesia submitted its updated Nationally Determined Contribution in 2022 with an unconditional emissions reduction target of 29% and a conditional target of 41% as compared to BAU scenarios. Further, the country is committed to achieve net-zero emissions by 2060. However, currently there are few policies that target climate mitigation and energy security at the city level, as most of the policies applicable to the city are at the provincial level.

Provinces are now required to develop a Regional Energy Plan to show how they can meet the national ambition for climate change mitigation reflected in the Nationally Determined Contribution commitments. The new clean energy bill (Regulation 45/2019) for Bali province—the first local regulation on clean energy—emphasizes zero energy buildings, and providing special tariffs to hotels, malls, and other commercial buildings that use electricity from clean sources. The Green Bali Plan with Net Zero Emission was announced in 2022. Under this plan, Bali aims to achieve net-zero emissions in 2045, which is faster than the national target of 2060.

A mapping of Indonesia's renewable energy potential has been carried out. Geothermal development had previously been explored, but environmental constraints along with limited technological capability have prevented further implementation. The government has secured a $380 million loan from the World Bank and the Asian Infrastructure Investment Bank to set up a pumped storage hydropower plant (in Upper Cisokan), which enables the country to have more renewable energy integrated into the national grid and provides a future option of sustainable energy storage.[2]

A large part of the energy consumed in the city is imported from the Java–Bali grid and currently, there are no plans to reduce the city's import dependence. However, the city government has started looking into renewable sources to improve Denpasar's energy security and develop the city as a sustainable tourism hub. With growing energy demand in the coming years, plans are being drawn up to set up a 20-megawatt (MW) waste-to-energy plant in collaboration with PT Siemens Indonesia.[3]

The local electricity company PLN Distribusi Bali is also exploring the economic feasibility of solar rooftop panels by setting up pilot projects of 30 kW and 10 kW.

[2] Umesh Ellichipuram. 2021. World Bank approves loan for hydropower plant in Indonesia. *Power Technology*. Press Release. 13 September.

[3] Power Technology. Power plant profile: Denpasar Waste-to-Energy Power Plant, Indonesia.

2. Energy Demand Outlook

This chapter details the potential future trajectories of energy demand in the city.

Total Final Energy Demand by Fuel

Oil is the major energy source, meeting 56% of the city's TFED in 2020, with gas and liquefied petroleum gas meeting around 20% each (Figure 5).

Figure 5 Share of Fuels in the Energy Demand of Denpasar in 2010 and 2020

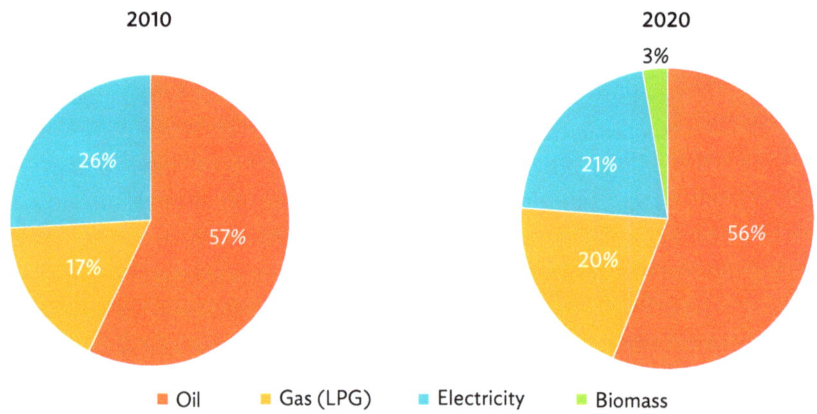

lpg = liquefied petroleum gas.

Source: Authors.

Business-as-Usual Scenario

Under the BAU scenario, the TFED reaches 2.026 Mtoe by 2050 with an annual growth rate of over 6%. Electricity and oil continue to be the primary fuels for meeting demand up to 2050, with an increasing use of electricity (Figure 6).

Low-Carbon Scenario

The pattern is not markedly different under the low-carbon scenario with oil and electricity dominating energy demand. However, in absolute terms, in 2050 TFED is around 10% lower than in the low-carbon scenario than in the BAU scenario and within electricity, off-grid supplies have a rapid growth of around 15% annually as compared with 7% for grid electricity (Figure 7). Oil demand is projected to grow annually at 6% and gas at 0.6%. By 2050, the largest share of TFED still comes from oil, with more than 51% share, followed by grid electricity at 39% and gas at 4%. The share of off-grid electricity reaches more than 6% in 2050, from almost negligible in 2030 (Figure 8).

Figure 6 **Fuel-Wise Final Energy Demand Under the Business-as-Usual Scenario (2020–2050)**

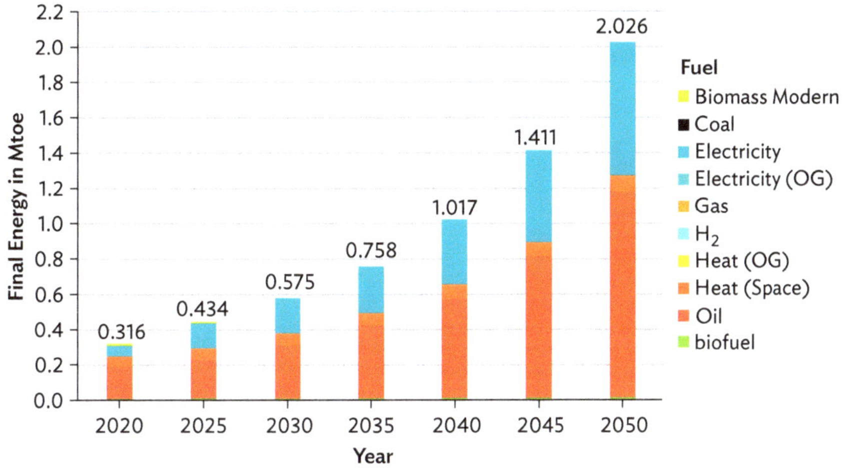

H_2 = green hydrogen, Mtoe= million tons of oil equivalent, OG = off-grid.

Source: Authors.

Figure 7 **Fuel-Wise Final Energy Demand Under Low-Carbon Scenario (2020–2050)**

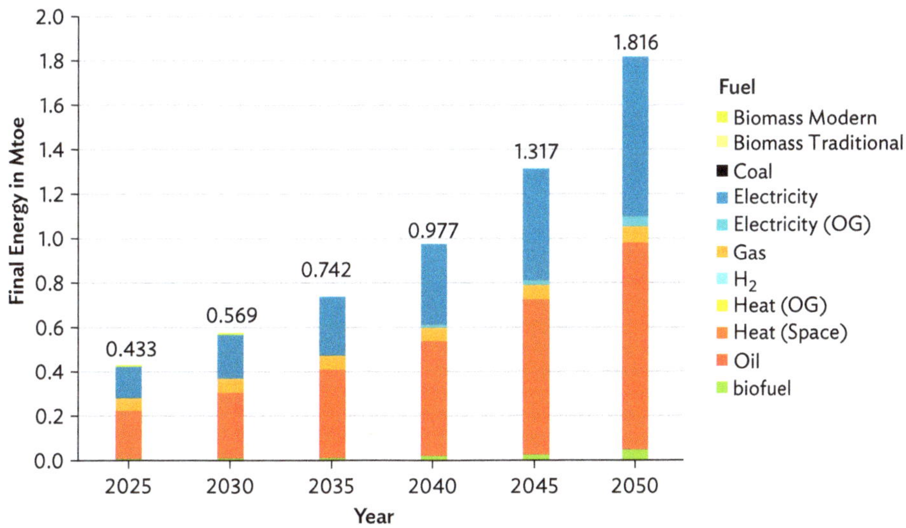

H_2 = green hydrogen, Mtoe= million tons of oil equivalent, OG = off-grid.

Source: Authors.

Figure 8 Fuel-Wise Share of Final Energy Demand Under Low-Carbon Scenario (2020–2050)

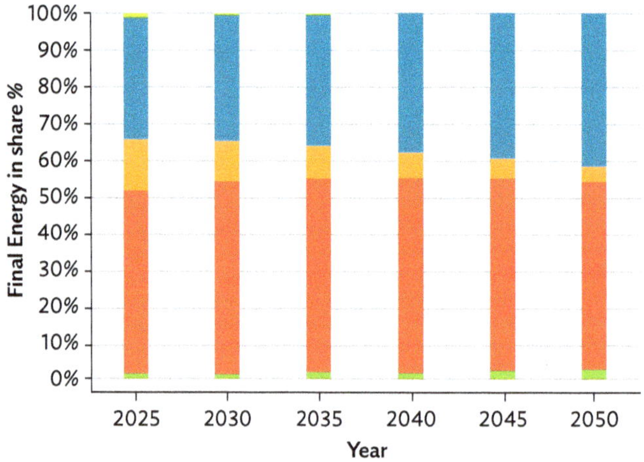

Source: Authors.

Total Final Energy Demand by Sector

In 2020, transport had the largest share in TFED at 55% followed by the residential (domestic) sector at 35%. The commercial sector accounts for 9% and because there is no large-scale industry in the city, the share of industrial sector is only 1% (Figure 9).

Figure 9 Share of Sector Energy Demand of Denpasar in 2010 and 2020

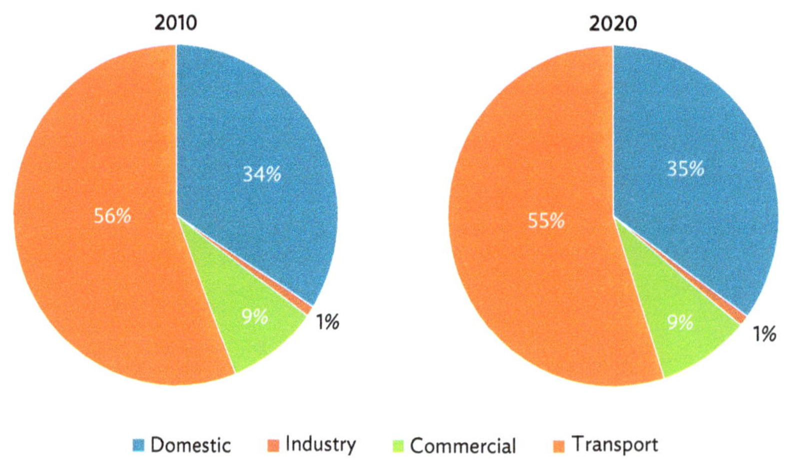

Source: Authors.

Business as-Usual Scenario

Under the BAU scenario, by share transport and residential continue to dominate (Figure 10). The share of transport, including both freight and passenger modes, reaches around 58% in 2050, while the share of residential is projected to grow steadily from 21% in 2030 to 23% in 2040 and 25% in 2050 (Figure 11). Industry's share remains very small due to the assumed continued absence of large industries in the city.

Figure 10 Total Final Energy Demand by Sector Under Business-as-Usual Scenario (2020–2050)

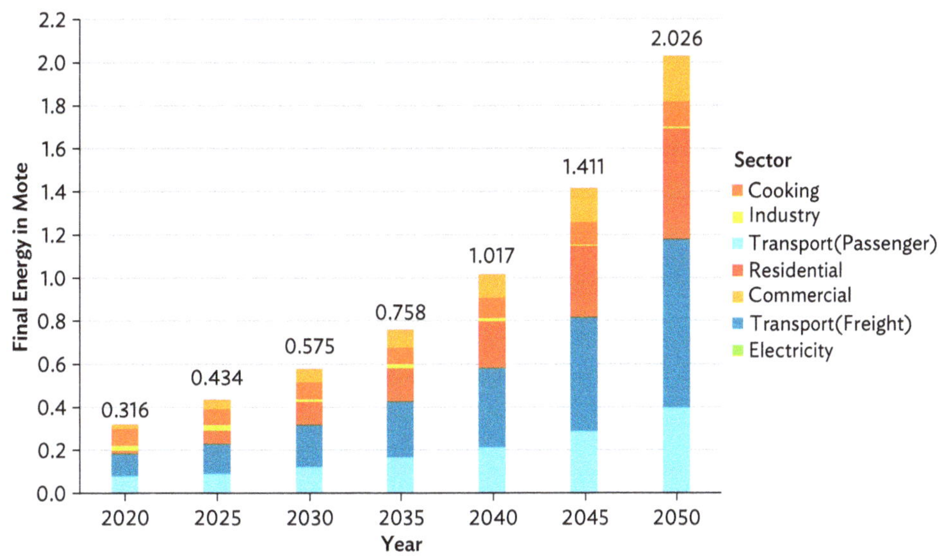

Mtoe = million tons of oil equivalent.

Source: Authors.

Figure 11 Sector-Wise Share of Final Energy Demand Under Low-Carbon Scenario (2025–2050)

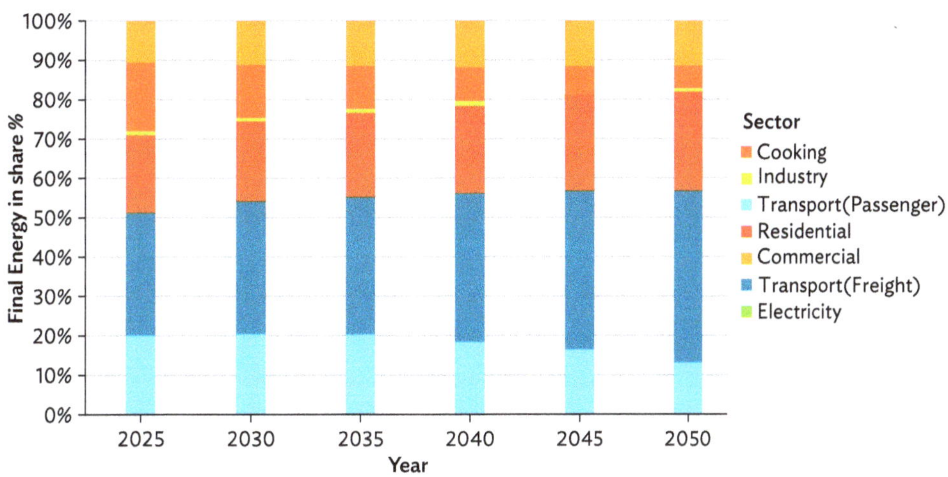

Source: Authors.

The fastest annual growth in energy consumption is expected in buildings (8%) while the slowest is in cooking (2%). Consequently, the share of residential cooking in TFED falls from 13.5% in 2030 to 5.4% in 2050. Even under the BAU there is expected to be a major shift away from biomass cooking with a decline of 75% in demand from biomass cooking (2020–2050). By 2050 roughly 84% of cooking demand is expected to be from gas stoves, with most of the remainder from electric cooking. As noted, transport is the major energy demand sector. Transport energy demand is projected to be almost 0.31 Mtoe in 2030 increasing to 1.17 Mtoe by 2050. Within transport, TFED by freight will be 0.19 Mtoe in 2030, increasing to 0.78 Mtoe in 2050, and taking 67% of total freight demand in 2050, with passenger transport taking the remainder. The BAU scenario includes only a very modest growth in electric vehicles. By 2050 electric two-wheelers comprise around 5% of all two-wheelers in the city, but electric cars only 1% of all cars. However, even under the BAU scenario all buses are taken to be electrified by 2030 (Table 4).

Table 4 Number of Electric Passenger Road Vehicles Under the Business-as-Usual Scenario

	2020	2025	2030	2035	2040	2045	2050
Two-wheelers	11,737	18,020	27,369	42,563	67,566	110,641	180,923
Cars	1,948	2,064	2,164	2,324	2,549	2,816	3,105
Buses	0	0	3,452	4,118	5,019	6,158	7,546

Source: Authors.

Low-Carbon Scenario

Under the more ambitious climate policy applied in this scenario, as noted by 2050 the TFED is 10% lower than in the BAU case. Transport and residential have the largest shares with similar annual growth rates of around 6%–7% over 2030–2050. Transport has a 56% share by 2050 and the share of industry remains negligible (Figure 11).

The total TFED by residential is about 0.19 Mtoe in 2030 and 0.3 Mtoe in 2050. The share of demand from residential cooking falls from 13% in 2030 to 6% in 2050, as there is a significant shift in the fuel mix, with a major decline in TFED for cooking from biomass stoves, which falls from 0.003 Mtoe in 2030 t0 0.001 Mtoe in 2050. Electricity demand for cooking rises from 0.009 Mtoe in 2030 to 0.029 Mtoe by 2050 (Figure 12). Residential demand for non-cooking applications like heating water and air cooling is 0.11 Mtoe in 2030 increasing to 0.47 Mtoe in 2050.

Transport is the other important user sector where there is a significant shift in terms of fuel use in the low-carbon scenario. The total TFED from transport is projected to be 0.31 Mtoe in 2030 increasing to about 1.02 Mtoe by 2050. Most freight is transported by road within the city boundaries in heavy duty vehicles and its energy demand grows at 7% annually, taking 76% of the total in 2050. Passenger transport demand grows at 4% and takes the remaining 24% in 2050.

In terms of TFED by various fuels and commodities there is projected to be a steady growth in energy demand from electricity-based, biofuel-based, and gas-based vehicles and away from fossil fuels (Figure 13 and Figure 14). In terms of passenger road traffic, the low-carbon scenario assumes a significant growth of electric vehicles, with a rising share of electricity in the fuel demand from road transport (Figure 15).[4]

[4] The fuel-switching interventions related to electric vehicles and biofuels in the road transport estimate have been applied just for the passenger segment and not the freight segment, because of lack of information on freight road traffic.

Figure 12 Residential Cooking Final Energy Demand Under Low-Carbon Scenario (2025–2050)

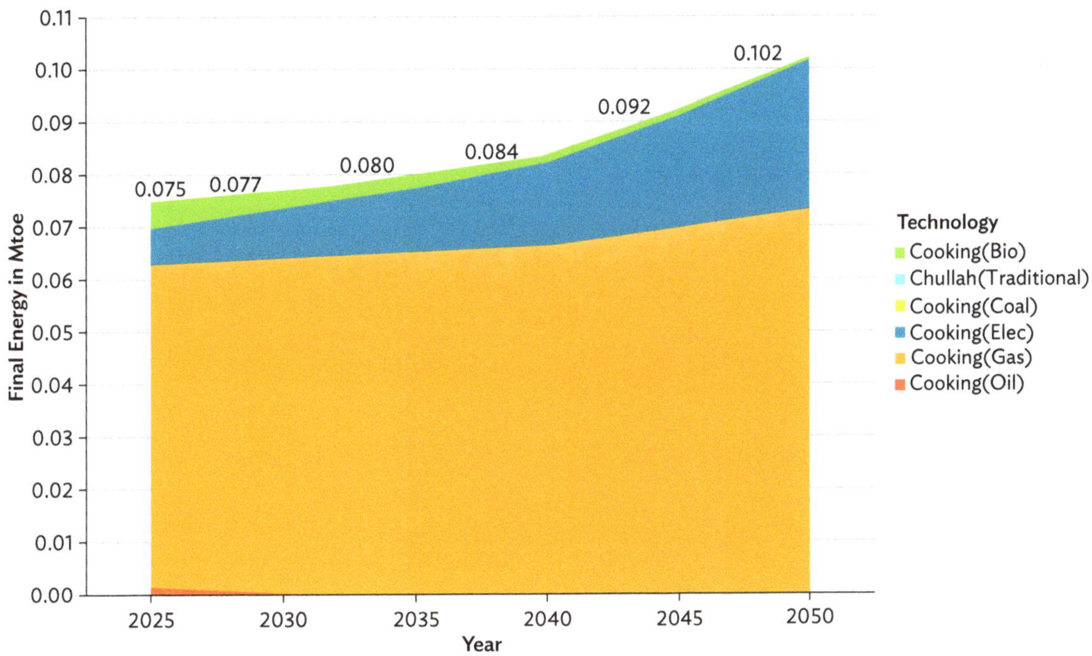

Elec = electricity, Mtoe = million tons of oil equivalent.

Source: Authors.

Figure 13 Transport (Freight Segment) Fuel-Wise Final Energy Demand Under Low-Carbon Scenario (2025–2050)

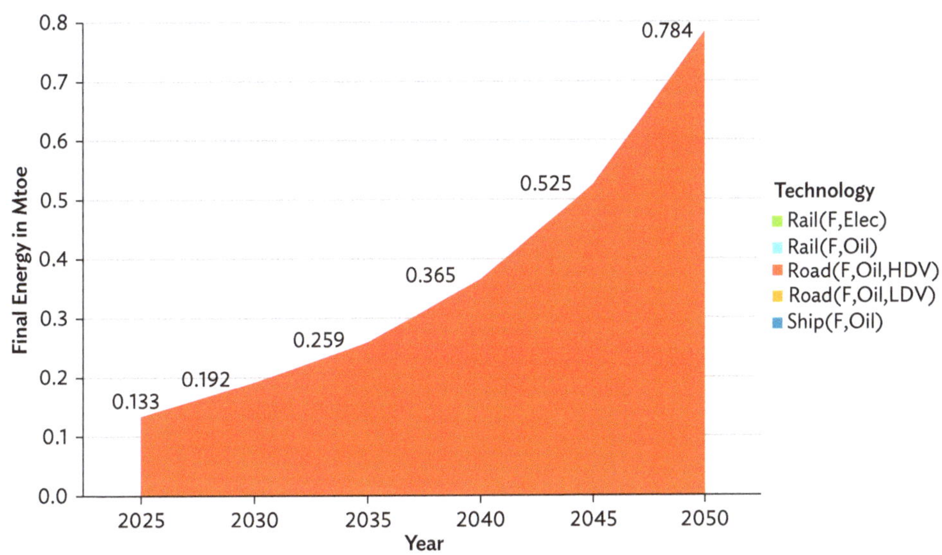

F= freight, Elec = electricity, HDV = heavy duty vehicle, LDV = light duty vehicle, Mtoe = million tons of oil equivalent.

Source: Authors.

Figure 14 Transport (Passenger Segment) Fuel-Wise Final Energy Demand Under Low-Carbon Scenario (2025-2050)

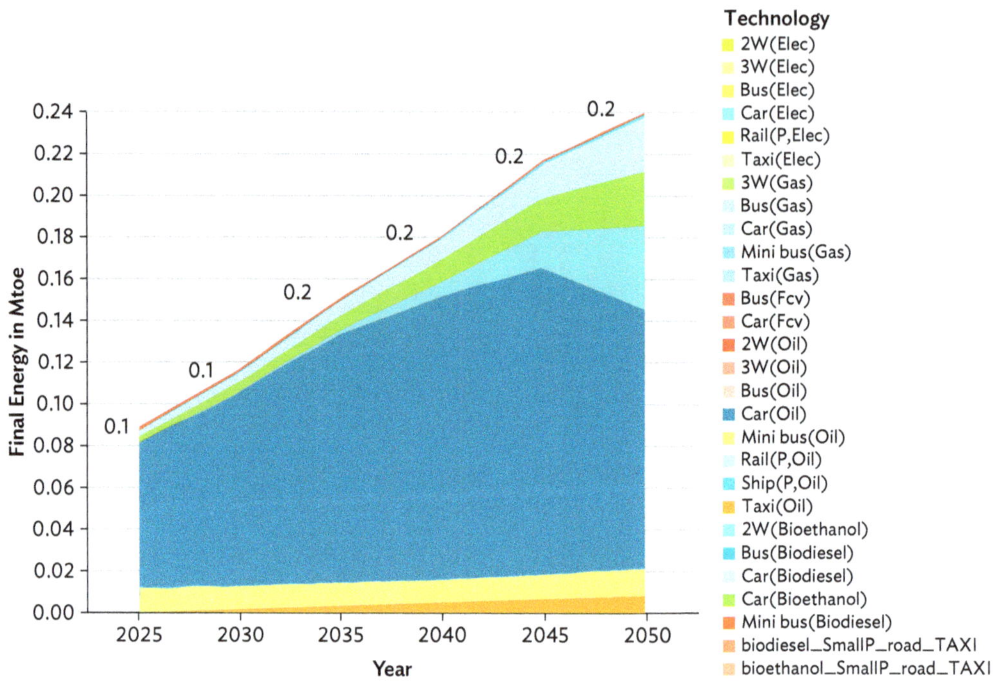

Elec = electricity, FCV = fuel cell vehicle, Mtoe = million tons of oil equivalent, SmallP_road_Taxi = small passenger road taxi

Source: Authors.

Figure 15 Passenger (Road Segment) Fuel-Wise Energy Demand Under Low-Carbon Scenario (2025-2050)

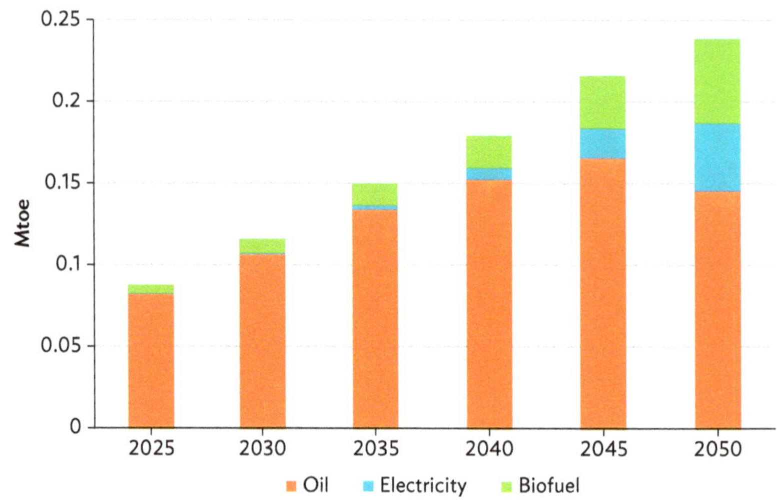

Mtoe = million tons of oil equivalent.

Source: Authors.

Table 5 shows the number of electric vehicles in Denpasar by type under the low-carbon scenario. The Bali Provincial Government has announced a target to electrify around 140,000 two-wheelers by 2026 and the estimates are consistent with these targets.[5] The number of electric two-wheelers is projected to grow at 17%, and electric cars at 18% between 2020 and 2050. By 2050, the aim is that two-wheelers are 50% and electric cars are 53% of the city market by 2050. All buses are taken to be electrified by 2030. These figures for electric vehicles far exceed those in the BAU, where only 3,105 electric cars are projected for the city in 2050, as compared with over 250,000 in the low-carbon scenario.

Table 5 Number of Electric Passenger Road Vehicles Under the Low-Carbon Scenario

	2020	2025	2030	2035	2040	2045	2050
Two-Wheelers	11,737	100,946	168,159	277,103	467,562	825,404	1,467,286
Cars	1,948	4,195	10,624	26,630	60,562	129,512	256,414
Buses	0	2,519	2,964	3,452	4,118	5,019	6,158

Source: Compiled by authors.

5 *Antara. 2023.* Bali targets use of 140 thousand electric motorcycles in 2026. 26 January.

3. Energy Supply Outlook

Primary Energy Supply

Oil has been the main source of energy for the city, although its share reduced significantly from 61% in 2010 to 50% in 2020 due to an increase in the share of grid-imported electricity, which reached 34% of TPES in 2020 (Figure 16). The share of gas (17% in 2020) was largely due to the use of liquefied petroleum gas in cooking and heating water.

Figure 16 Primary Energy Mix for Denpasar in 2010 and 2020

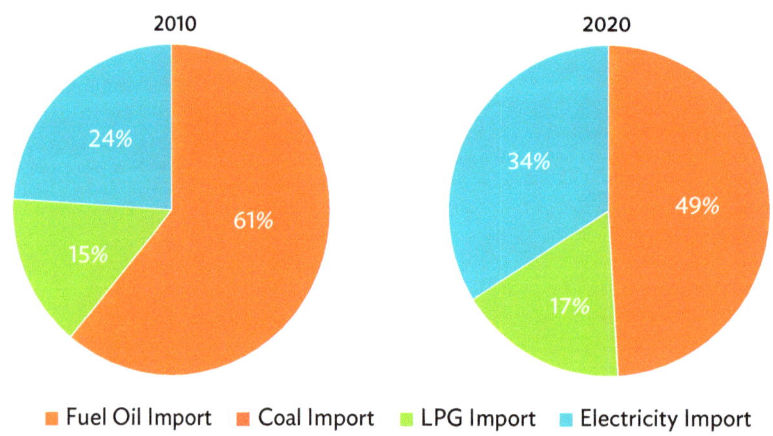

LPG = liquefied petroleum gas.

Source: Authors.

Business-as-Usual Scenario

In the BAU scenario, the TPES grows at about 6.5% annually between 2030 and 2050, reaching 2.3 Mtoe by 2050. Most of the TPES (99%) comes from outside the city with energy imports (defined as any energy not generated from the city itself) dominated by oil and electricity (Figure 17). Energy imports are projected to grow from 0.64 Mtoe in 2030 to 2.29 Mtoe in 2050, an annual rate of 7%. Import share remains constant at around 99% between 2030 and 2050 (Figure 18). The small amount of domestically supplied primary energy from the city area is largely from solar.

Figure 17 Primary Energy Imports by Source in Denpasar Under Business-as-Usual Scenario (2020–2050)

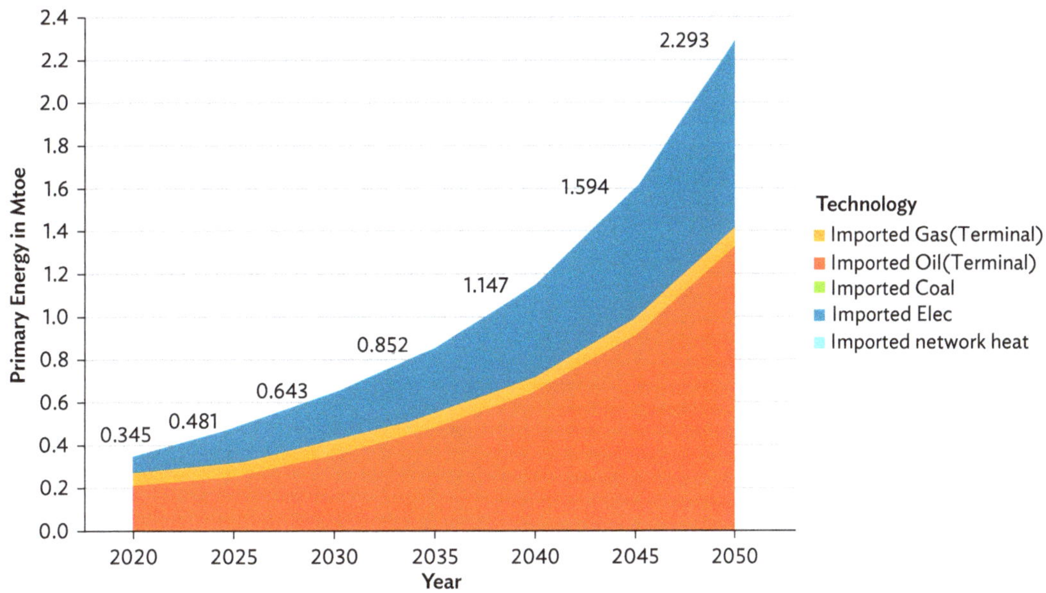

Elec = electricity, Mtoe = million tons of oil equivalent.

Source: Authors.

Figure 18 Fuel-Wise Domestic Primary Energy Supply for Denpasar Under Business-as-Usual Scenario (2020–2050)

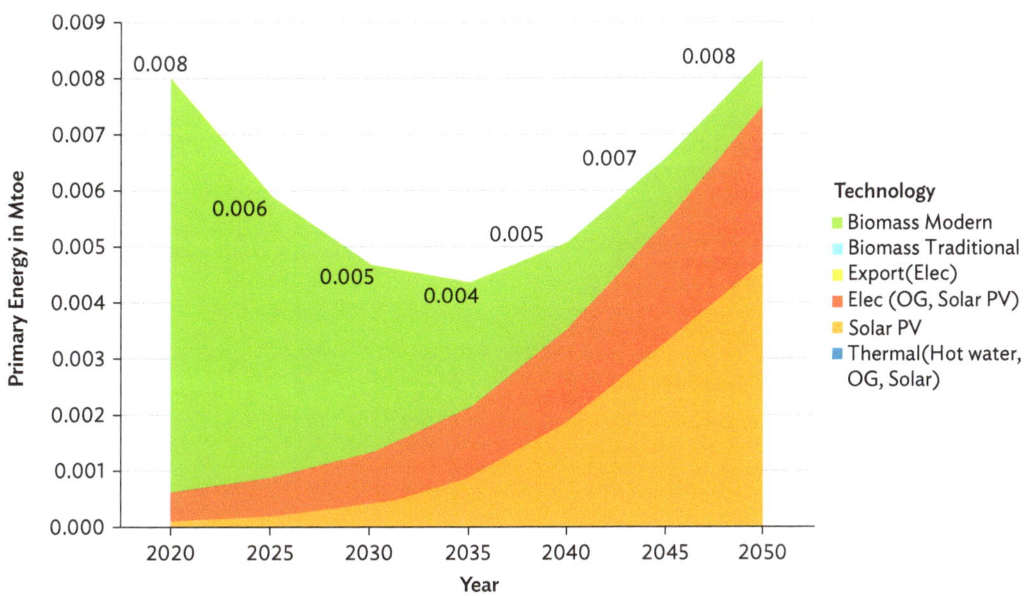

Elec = electricity, OG = off-grid, Mtoe = million tons of oil equivalent, PV = photovoltaic.

Source: Authors.

Low-Carbon Scenario

Under the low-carbon scenario, TPES reaches 0.64 Mtoe by 2030, increasing to 2.06 Mtoe by 2050, an annual growth of about 6%. By 2050 TPES is 13% below that in the BAU scenario, as a result of the assumed adoption of energy efficiency measures. As in the BAU case, energy imports from outside the city dominate TPES and the key energy fuels imported are oil and electricity. Electricity imports increase from 0.3 gigawatt (GW) in 2030 to 1.1 GW in 2050 (Figure 19).

Figure 19 Total Primary Energy Supply by Imports Under Low-Carbon Scenario (2020–2050)

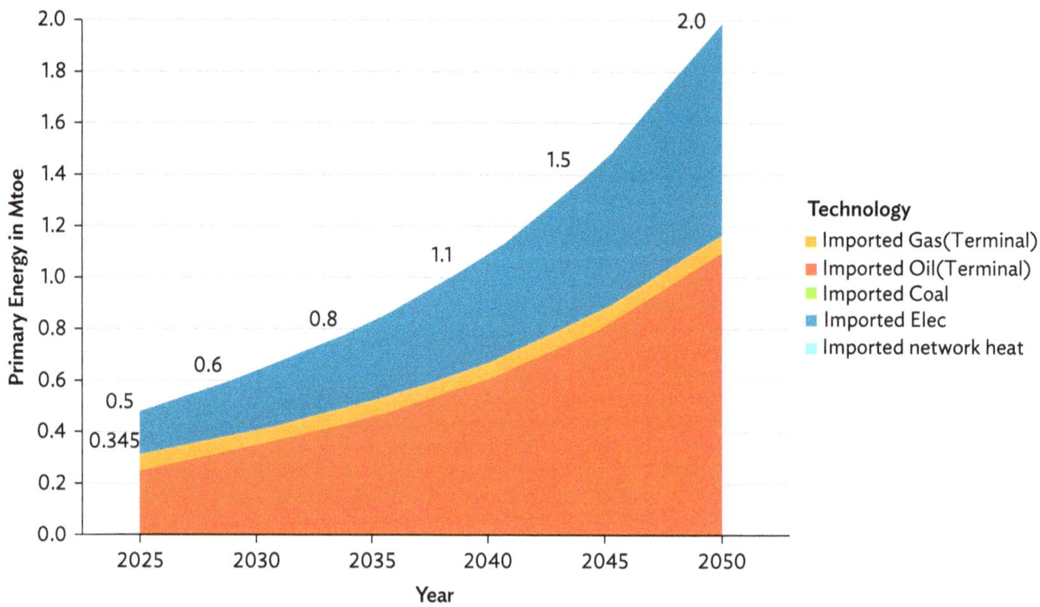

Elec = electricity, Mtoe = million tons of oil equivalent.

Source: Authors.

Domestic primary energy supply remains limited but increases far more rapidly than in the BAU scenario. It is projected to rise from 0.010 Mtoe in 2030 to 0.067 Mtoe by 2050 (as compared with 0.008 Mtoe in in 2050 in the BAU case), with an annual growth rate of 11% for the period 2030 to 2050. Close to 90% of domestic energy would be sourced from solar PV (grid and off-grid) by 2050. Also, use of biomass for cooking would be replaced by use of biofuels in transport (Figure 20).

Figure 21 gives the cumulative TPES by various fuels and commodities between 2025 and 2050 under the low-carbon scenario. Oil is the single largest fuel commodity taking 53% of the TPES in 2030 and 52% in 2050. The share of electricity rises from around 35% in 2030 to 40% in 2050, while that of solar energy increases from 0.5% in 2030 to 2.9% in 2050.

Both scenarios allow for a rising demand for energy in response to income and population growth; however under the low-carbon pathway, demand and the primary energy required to meet it are lower, due to the policy initiatives and adoption of energy-efficient technologies and processes that are assumed for the low-carbon scenario.

Figure 20 **Primary Energy Supply by Domestic Fuel Sources Under Low-Carbon Scenario (2020–2050)**

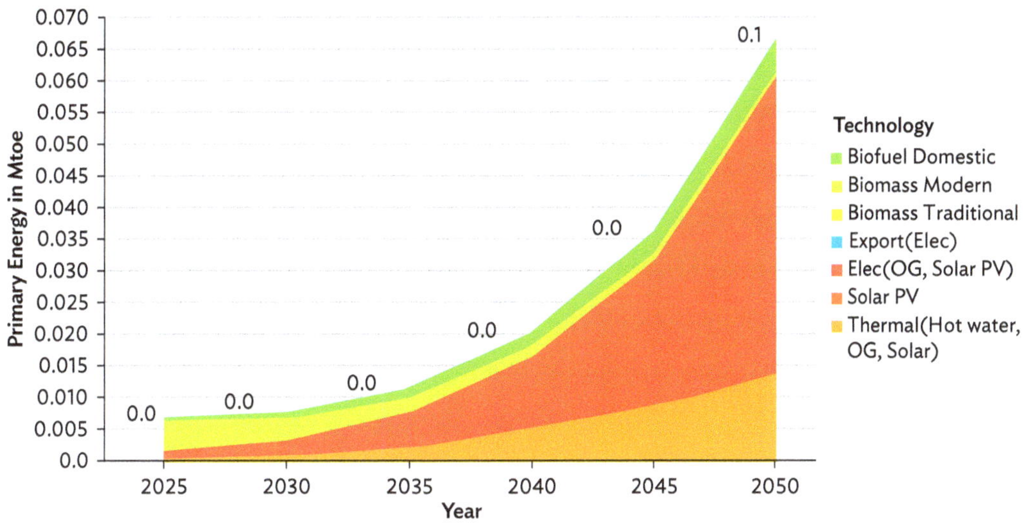

Elec = electricity, Mtoe = million tons of oil equivalent, OG = off-grid, PV = photovoltaic.

Source: Authors.

Figure 21 **Share of Total Primary Energy Supply by Fuel Under Low-Carbon Scenario (2020–2050)**

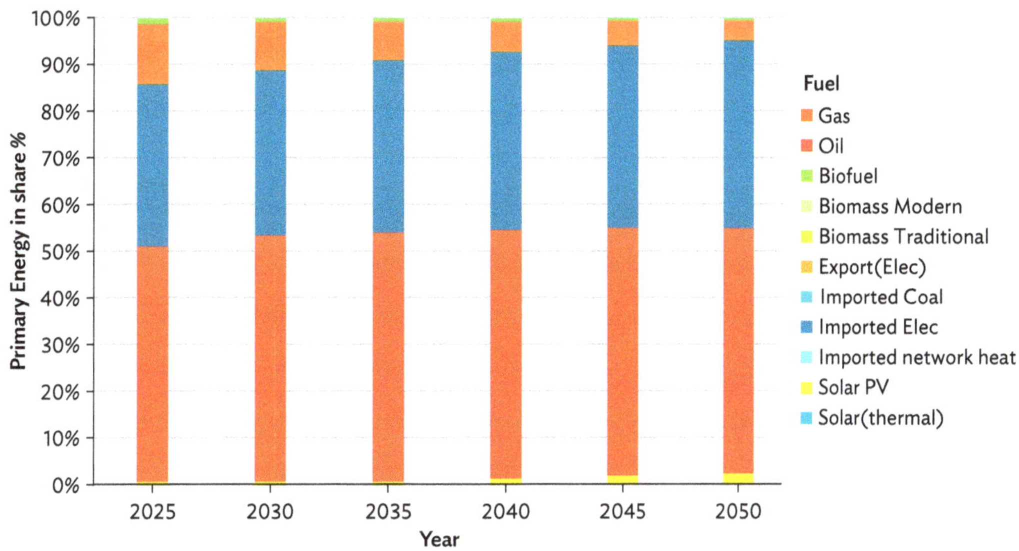

Elec = electricity, PV = photovoltaic.

Source: Authors.

Electricity Generation Outlook

Natural gas was the only fuel used for electricity generation in 2020, as the city has oil-based power plants, which have been upgraded for use as dual plants (gas and diesel). Electricity generation from oil-fired plants was significant in the past, reaching 30% in 2015, however after this date gas-based power generation has provided virtually all the domestic supply (Figure 22).

Figure 22 Electricity Generation by Fuel Type (2010–2020)

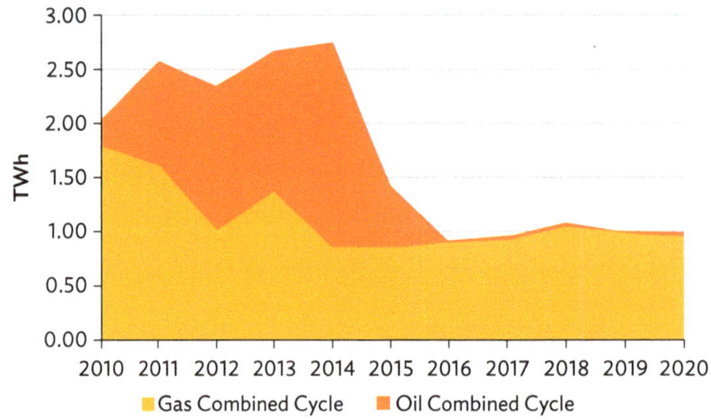

Twh = terawatt-hour.

Source: Authors.

Business-as-Usual Scenario

The total installed capacity of the grid-connected power plants (gas and diesel-firing) in the region is currently 552 MW and is expected to stay stable, while the maximum capacity that the grid can deliver (local + imported) is 1,033 MW. Under the BAU scenario total electricity generation required for the city grows at 7% annually (2030–2050), from 2.63-terawatt-hours (TWh) in 2030, to 5.3 TWh in 2040 and 10.5 TWh in 2050, with the majority supplied from the national Java–Bali electricity grid (accounted here as electricity import). Domestic supply from the city itself comes from solar PV plants and grows from 3.2 MW in 2030 to 14.1 MW in 2040 and 34.3 GW in 2050.[6] This is a rapid growth of 13% annually, but from a very low base (Figure 23).

Low-Carbon Scenario

Under the low-carbon scenario, electricity generation reaches 2.7 TWh by 2030 rising to 10.3 TWh by 2050, a growth rate of 7% (Figure 24). The cumulative quantum of electricity imported increases to 0.3 GW in 2030 and 1.1 GW in 2050. As in the BAU scenario, capacity from solar PV-based plants located in the city is expected to account for a small share of the total electricity mix; however, the number and capacity of solar installations, a majority of which are off-grid, are much higher in the low-carbon than in the BAU scenario (410.8 MW as opposed to 44.6 MW in 2050) (Figure 25).

[6] Although this is a very small share of renewables it should be noted that some of the grid electricity will be from renewable sources.

Figure 23 Electricity Generation by Source Under Business-as-Usual Scenario (2020–2050)

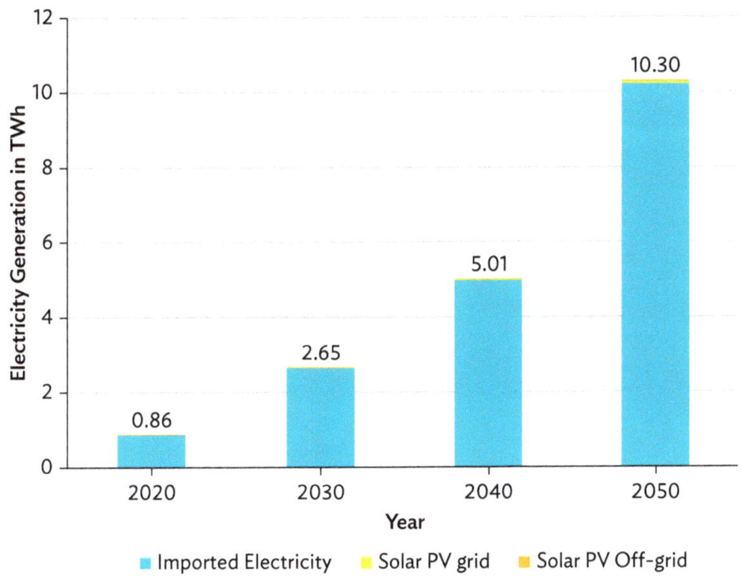

PV = photovoltaic, TWh = terrawatt-hour.

Source: Authors.

Figure 24 Electricity Generation by Source Under Low-Carbon Scenario (2020–2050)

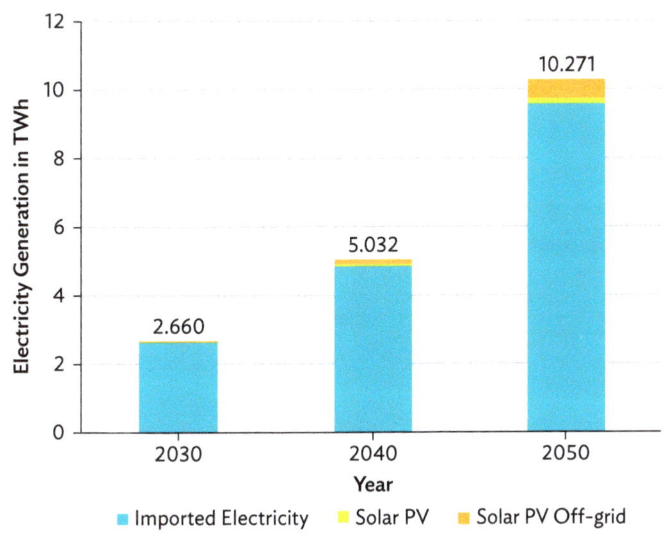

PV = photovoltaic, TWh = terrawatt-hour.

Source: Authors.

Figure 25 Cumulative Installed Electricity Capacity by Source Under Low-Carbon Scenario (2020–2050)

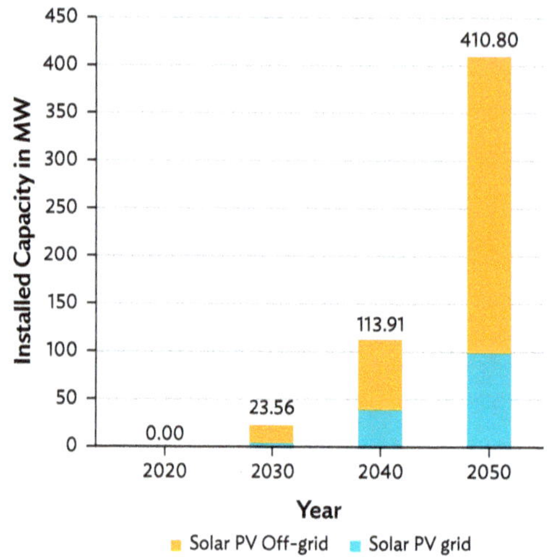

MW = megawatt, PV = photovoltaic.

Source: Authors.

4. Emissions Outlook

Emissions Projections

The model applies the energy consumption of the sectors covered and the emissions factors associated with the energy commodities they use to estimate city-level GHG emissions from both the demand and supply sides in the period up to 2050.

Business-as-Usual Scenario

The total GHG emissions (supply side and demand side) in the city under the BAU scenario are 1.37 million tons of carbon dioxide equivalent ($MtCO_2e$) in 2030 and 4.8 $MtCO_2e$ in 2050, an annual total emissions growth of 6% (2030–2050). Most emissions (approximately 90%) come from the demand sectors. Demand sectors in total are projected to emit 1.2 $MtCO_2e$/year in 2030 and 4.3 $MtCO_2e$/year in 2050. Transport in the city is by far the greatest emitter and will cause 87% of demand-side emissions in 2030, 92% in 2040, and 95% in 2050. Residential cooking sees a decrease in its share from 13% in 2030 to 8% in 2040 and 5% in 2050, even under the BAU, due to the shift to more modern forms of cooking (Figure 26).

Figure 26 Total Demand-Side Greenhouse Gas Emissions Under Business-as-Usual Scenario (2020–2050)

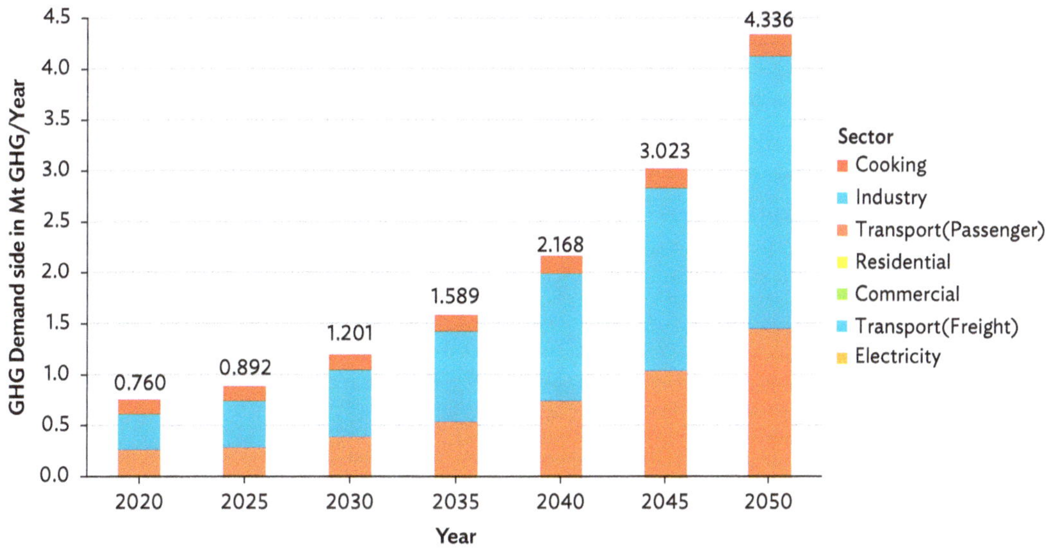

GHG = greenhouse gas, Mt = million tons.

Source: Authors.

Emissions from the supply side are much lower than from the demand side, increasing from 0.123 $MtCO_2e$/year in 2030 to 0.223 $MtCO_2e$/year in 2040 and 0.454 $MtCO_2e$/year in 2050, a growth rate of about 6.5%. Virtually all these come from the use of oil as a fuel (Figure 27).

Figure 27 Total Supply-Side Greenhouse Gas Emissions Under Business-as-Usual Scenario (2020–2050)

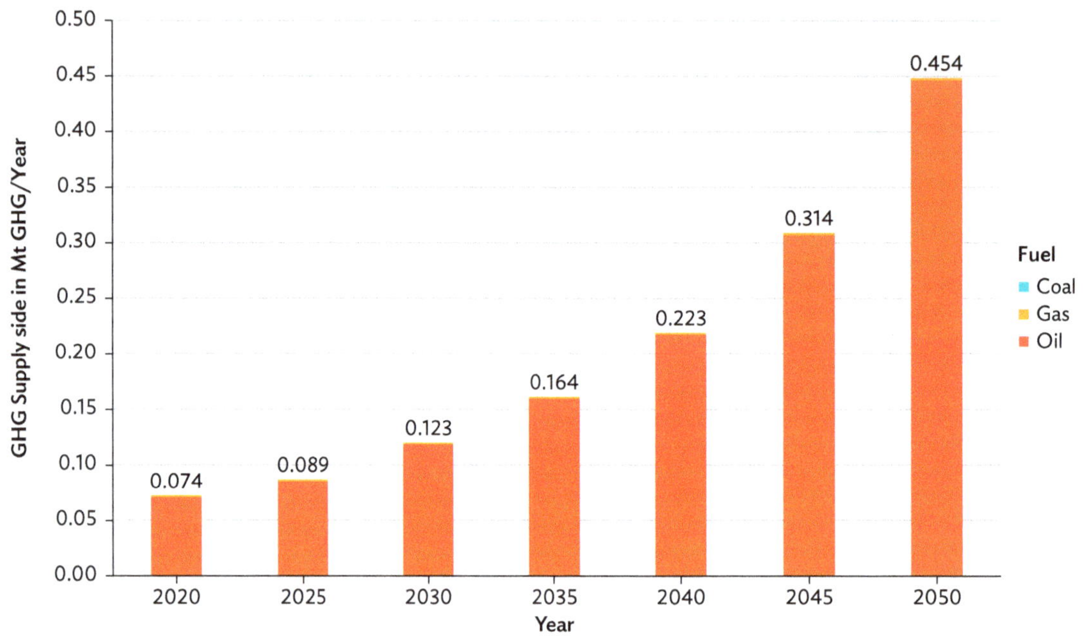

GHG = greenhouse gas, Mt = million tons.

Source: Authors.

Low-Carbon Scenario

A key objective of climate policy is to reduce emissions. Under the low-carbon scenario total GHG emissions in the city are projected to grow from 1.3 $MtCO_2e$ in 2030 to 3.95 $MtCO_2e$ in 2050, which is still an emissions annual growth of a little below 6%. However, significantly, by 2050 total emissions are 18% below the BAU figure. As in the BAU scenario, demand-side emissions dominate the total, rising from 1.2 $MtCO_2e$ in 2030 to 3.6 $MtCO_2e$ by 2050. Figure 28 shows the sector mix of demand-side GHG emissions between 2025–2050, which as before are dominated by transport and to a lesser extent residential, and in total are roughly 18% below the BAU figure by 2050.

Total supply-side GHG emissions are again much lower than those on the demand side, reaching 0.122 $MtCO_2e$ in 2030 and 0.391 $MtCO_2e$ in 2050. By 2050, these are 14% lower than in the BAU scenario. Figure 29 shows the fuel distribution for supply-side GHG emissions generation 2025–2050, which remains largely from use of oil.

Figure 28 Demand-Side Greenhouse Gas Emissions Under Low-Carbon Scenario (2020–2050)

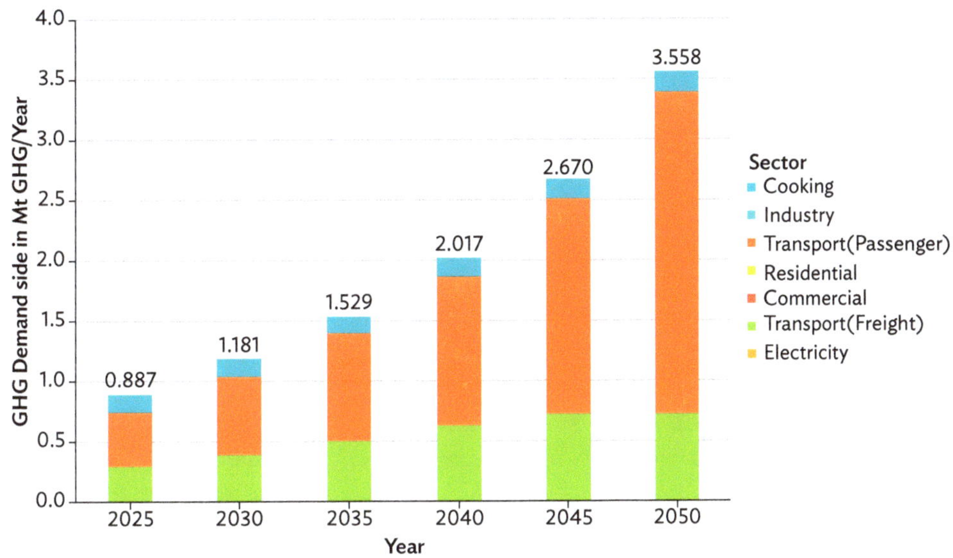

GHG = greenhouse gas, Mt = million tons.

Source: Authors.

Figure 29 Supply-Side Greenhouse Gas Emissions Under Low-Carbon Scenario (2020–2050)

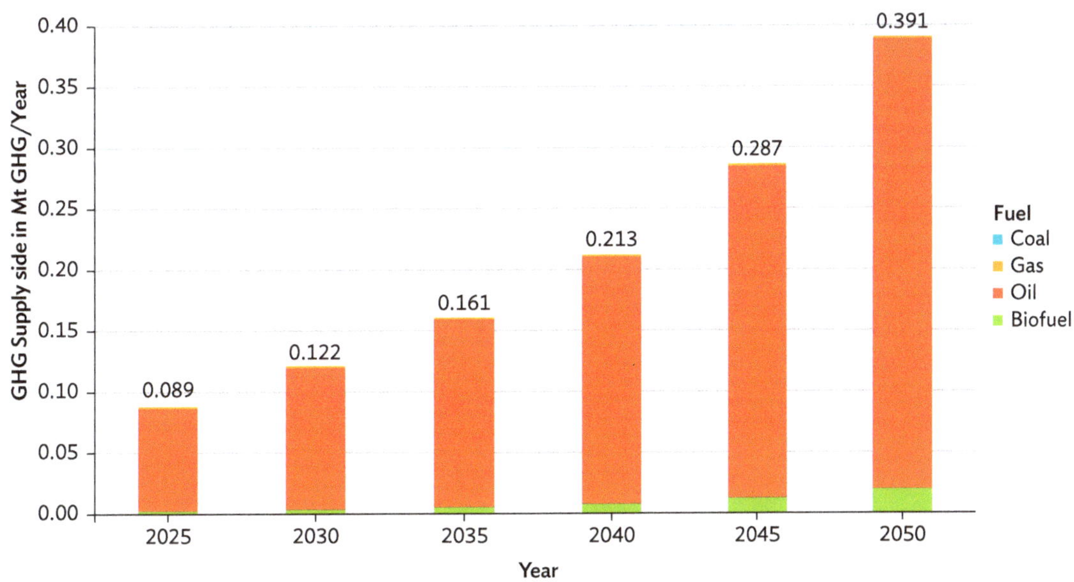

GHG = greenhouse gas, Mt = million tons.

Source: Authors.

Emission Mitigation Potential

Based on data collected for energy consumption in the different sectors, the low-carbon scenario identifies the key areas with a significant potential for emissions mitigation.[7] These are summarized in Table 6. Change within the residential sector is found to have by far the greatest impact.

- **Residential sector.** Changing cooking from traditional gas and biomass to electricity will increase efficiency and help reduce emissions. Installation of efficient cooling appliances such as inverter air-conditioners and high-efficiency chillers will help reduce total energy demand, while the installation of rooftop solar PV will reduce dependence on grid supply and further reduce emissions. With an energy transition in cooking and the use of efficient appliances, combined with improvements in buildings, the low-carbon scenario estimates a total of 9.8 $MtCO_2$ emissions mitigation can be achieved from the residential sector between 2020 to 2050, including emissions mitigation of 6 $MtCO_2$ from cooking alone. Hence improvements in cooking practice and use of low-carbon-intensive cooking appliances are critical.
- **Energy supply.** The analysis in the low-carbon scenario estimates that using utility scale and rooftop solar PV-based power generation to complement imports from the grid, an emissions mitigation of 0.92 $MtCO_2$ can be achieved between 2020 and 2050. This includes 0.34 $MtCO_2$ emissions mitigation from rooftop solar PV power plants on residential and commercial buildings.
- **Transport sector.** Transitioning from petrol and diesel-powered to electric vehicles is estimated to mitigate 1.4 $MtCO_2$ emissions between 2020 to 2050.
- **Commercial sector.** Installation of rooftop solar PV units in this sector will reduce dependence on grid supply and support the transition to a cleaner energy supply. The low-carbon scenario estimates emissions mitigation of 3.1 $MtCO_2$ from the commercial sector 2020 and 2050.

Table 6 Emissions Mitigation in Low-Carbon Scenario (by Sector)

Sector	Emissions Mitigation $MtCO_2$ (Low-Carbon Scenario 2020–2050)
Energy supply	0.92
Residential	9.8
Of which cooking	6.0
Transport	1.4
Commercial	3.1

$MtCO_2$ = million tons of carbon dioxide.

Source: Compiled by authors.

Some of the technologies used in the analysis of the low-carbon scenario to achieve this projected mitigation of emissions are summarized in Table 7.

In addition to the technologies in Table 6, various low-carbon technologies, which are not included in the low-carbon scenario due to their high cost at present, might be considered in the future if their cost reduces sufficiently. These options include the use of hydrogen in transport and battery energy storage for high renewable energy integration.

[7] In the model mitigation potential is calculated in a comparison with the most polluting coal alternative.

Table 7 Emissions Reducing Technologies in Low-Carbon Scenario

Technology	Description	Mitigation Potential	Existing Policy and Measures	Country Context
Electric vehicles	Electricity-based technology that would become an alternative to internal combustion vehicles (ICE)	The penetration of electric vehicles is expected to have mitigation potential of 1.4 million tons of carbon dioxide equivalent ($MtCO_2e$)	Presidential Decree 55/2019 (Acceleration Program for Battery-powered Electric Vehicles for Road Transportation)	The Government of Indonesia expects 400,000 electric vehicles to be under operation by 2025.[a] Also, with large nickel laterite ore reserves, there are plans to target battery production domestically with production of electric vehicles.
High- efficiency residential and commercial heating, ventilation, and air-conditioning (HVAC)	Use of inverter technology and fixed speed air-conditioners for better performance of the air-conditioners along with common metric and standards and labelling program to increase savings and quality	Efficient appliance penetration has mitigation potential of 7 $MtCO_2e$ from the building sector	Minister of Energy and Mineral Resources Regulation No. 57/2017 Joint Credit Mechanism, a bilateral partnership between Indonesia and Japan that supports Indonesian companies to purchase high efficiency HVAC technologies from Japan (Sustainable Energy for All, 2019)	Around 223 million air- conditioners were operative in Indonesia as of 2019. In 2016, the share of electricity consumption for cooling purposes during peak loads reached 15%. This is expected to increase up to 40% by 2050, without efficient cooling systems. In 2021, The Directorate of Energy Conservation drafted new Minimum Energy Performance Standards (MEPS) and labelling policies for chillers and refrigerated display cabinets. Both appliances provide an opportunity for efficiency improvements and emissions reduction. By 2030, the policies have the potential to prevent 10.5 million tons of carbon dioxide equivalent ($MtCO_2e$) and reduce energy consumption by 11,725 gigawatt-hours.[b]

continued on next page

Table 7 *continued*

Technology	Description	Mitigation Potential	Existing Policy and Measures	Country Context
Decentralized generation and mini grids	A mini-grid is defined as a system having a renewable energy-based electricity generator (with capacity of 10 kilowatts and above), and supplying electricity to a target set of consumers through a public distribution network	Extension of decentralized power generation and off-grid system has a mitigation potential of 0.33 $MtCO_2e$	Ministry of Industry (MOI) Regulation No. 54/2012 MOI Regulation No. 5/2017 Minister of Energy and Mineral Resources (MEMR) Regulation No.38/2016	
Solar photovoltaic (PV) grid	PV panels or arrays are connected to the utility grid through a power inverter unit allowing them to operate in parallel with the electric utility grid. Extra power generated can be stored in batteries or returned to the grid.	Grid-based solar power generation is expected to have mitigation potential of 0.6 $MtCO_2e$	National Energy Policy 2014 National Energy Plan (RUEN-2017) Minister of Energy and Mineral Resources Regulation No. 49 (2018) Regulation No. 13 (2019) Regulation No. 16 (2019)	

[a] Patunru, A.. 2023. Why Indonesia wants Australia's help to supply world with EVs and batteries. https://www.pv-magazine-australia.com/2023/07/06/why-indonesia-wants-australias-help-to-supply-world-with-evs-and-batteries/ (23 December 2023)

[b] CLASP. 2022. Indonesian Cooling Appliance Policies to Avert 10.5 Mt CO_2 by 2030.

Source: Compiled by authors.

5. Investment Requirements and Recommendations

Requirements for a Low-Carbon Transition

Delivery of the emissions reduction projected in the low-carbon scenario requires both substantial additional investment and significant attitudinal change among city inhabitants. The city authorities and the federal government have an important role in the implementation of the range of low-carbon interventions, but the support of other stakeholders is needed. Key changes required for the clean energy transition in the city are as follows:

- **Energy-efficient appliances.** With the increasing energy demand from the building sector, there is a need to change consumer behavior toward energy-efficient appliances. This could involve bulk procurement by the city authorities or electricity distribution companies as a means of lowering prices for consumers, better information campaigns on their benefits, as well as cooperation with manufacturers to promote their use.
- **Electric vehicles.** Electrification of the transport sector requires investment in several areas, including development of charging infrastructure, improved after-sales service, and adapting the grid to allow additional electricity provision to the transport sector.
- **Solar photovoltaic power plants.** With growing electricity demand across the city and the national focus on a clean energy transition, the energy potential of solar PV power plants need to be utilized. These power plants can range from large-scale grid-tied power plants, rooftop solar PV systems, and off-grid solar PV systems. However, consumers need to be encouraged to set up solar rooftop systems, either financially or through information campaigns.
- **Electricity grid.** The electricity grid will need to be modernized to cope with growing electricity demand, expected electrification in the transport and residential cooking sectors, and increasing generation through solar PV systems. It needs to be flexible enough to evacuate the electricity generated through solar PV and to cater for fluctuating energy demand.

Investment Analysis

As with other aspects of the analysis the model compares energy investment required to support projected income and population growth without a major shift toward a low-carbon pathway (the BAU scenario) with investment needs allowing for such a shift (the low-carbon scenario). This chapter identifies the investment required for different sectors and power technologies under both scenarios. The numbers give cost for the installation of additional capacity, over that already in the system, averaged on an annual basis in constant 2010 US dollars. As these are constant price estimates, current price figures will be much higher.

Business-as-Usual Scenario

Total additional investment in the energy system under the BAU is estimated to be around $0.42 billion (2020–2030), $0.86 billion (2030–2040), and $1.78 billion (2040–2050). The majority of this is in transport and to some extent in residential. Supply-side investment is modest and mostly in relation to solar generation (grid-connected and off-grid). Table 8 shows the distribution of energy system investments under BAU scenario from 2025 to 2050.

Table 8 Investment Requirement Under Business-as-Usual Scenario for Denpasar ($ million 2010)

Sectors	2025	2030	2035	2040	2045	2050
Biofuels	0.02	0.02	0.03	0.02	0.04	0.05
Gas	0.00	0.11	1.76	0.13	0.21	0.23
Solar PV	0.39	0.73	1.27	2.65	3.33	2.96
Electricity (off-grid solar PV)	1.92	2.07	2.42	6.79	4.42	5.18
Electricity Grid	0.03	0.07	0.15	0.33	0.45	0.44
Cooking Sector	1.69	1.56	1.84	2.22	0.41	1.11
Residential Sector	16.62	22.92	32.96	48.67	70.92	114.65
Transport Sector	213.15	137.36	359.39	366.74	679.23	834.89
Commercial Sector	13.99	4.99	20.25	13.76	32.22	29.94
	0.17	0.55	0.66	0.87	1.35	1.97

PV = photovoltaic.

Source: Compiled by authors.

Low-Carbon Scenario

Total investment requirement under the low-carbon scenario is estimated to be around $0.43 billion (2020–2030), $0.95 billion (2030–2040), and $2.16 billion (2040–2050). Table 9 shows the distribution of this investment from 2025 to 2050.

Table 9 Investment Requirement Under Low-Carbon Scenario for Denpasar ($ million 2010)

Sectors	2025	2030	2035	2040	2045	2050
Biofuels	0.04	0.06	0.10	0.13	0.24	0.38
Gas	0.00	0.06	1.68	0.03	0.09	0.10
Solar PV	0.79	1.73	3.63	8.02	8.22	13.07
Electricity (off-grid solar PV)	7.22	12.91	25.24	50.34	81.85	160.55
Electricity Grid	0.08	0.17	0.41	1.01	1.03	1.66
Cooking Sector	1.72	1.56	1.85	2.30	0.49	1.51
Residential Sector	13.53	21.43	33.57	59.55	103.85	199.44
Transport Sector	213.25	137.84	359.90	368.91	684.16	845.69
Commercial Sector	12.35	4.37	17.88	12.15	28.45	26.44
Imports	0.17	0.45	0.50	0.63	0.95	1.28

PV = photovoltaic.

Source: Compiled by authors.

In total investment (2020–2050) under the BAU is $3.06 billion and under the low-carbon scenario it is 16% higher at $3.54 billion (in 2010 prices). Investment is 21% higher in the last decade of the period (2040–2050). In the low-carbon scenario the majority of the investment is projected to be in transport for the spread of low-carbon options like electric vehicles and biofuel blending and the associated infrastructure and in residential for deployment of more efficient appliances, particularly a district cooling system, with centrifugal chillers. Investment in centrifugal chillers alone is 70% of residential investment in 2050. On the supply side, relative to the BAU case, higher investment is projected for biofuels, solar PV (grid-connected), electricity (off-grid), and electricity (grid). The higher investment for grid electricity is required for upgrading the distribution system to cope with increased electrification resulting from greater e-mobility and electric cooking. Investment to support imports under the low-carbon scenario is lower, as a result of projected higher generation from domestic renewables.

Recommendations

Denpasar City, the capital of Bali Province, is a widely renowned tourist hub. Tourism therefore has a significant impact on the city's economy and energy needs, with wide seasonal variations. Sustainable tourism will be an essential element for the low-carbon energy transition and hence encouraging sustainable tourism will be critical for the city economy, its people, and the environment.

From the analysis of the low-carbon scenario, key steps for the city's decarbonization are the uptake of renewable energy, decarbonization of the transport sector, the promotion of efficient cooling in buildings and a shift away from traditional forms of cooking. Interventions to support these require involvement from both national and local government and the private sector, while power sector interventions, such as scaling up deployment of solar PV power plants, are mainly influenced by national policies.

Electricity is a significant part of energy demand in all cities. In addition, electrification of end-use demand also requires simultaneous efforts toward decarbonization of power generation. Decarbonizing electricity requires greater uptake of renewable energy. Solar rooftops are an effective way to decarbonize the grid in urban areas, where land is a constraint. However, so far in Indonesia, obstacles to further capacity installation from licensing processes and restrictions on selling excess electricity have limited investments in rooftop solar.[8]

While the deployment of large-scale solar is driven by investment by the government and large private companies, small-scale decentralized systems can be deployed by individuals in the form of solar PV rooftop systems. While various national level programs support their expansion, the city administration can promote their implementation more effectively by providing information on vendors of various kinds of associated services, fast-tracking their approval, and promoting best practices and demonstration projects.

Demand for energy for cooling is also important in cities and is ever increasing in Indonesia. Efficient cooling is a promising solution to reduce electricity demand. Innovative business models such as a city-level demand aggregation and a bulk procurement model can also be considered to reduce costs for consumers by aggregating the demand for efficient technologies and appliances and offering an integrated solution (e.g., a solar rooftop and efficient air-conditioning package). Another measure could be the implementation of efficient air-conditioning systems at the district level, for example by identifying buildings to be developed as a pilot project to showcase the technology and its benefits, particularly for new areas of the city.

8 Asian Development Bank. 2023. Proposed Programmatic Approach and Policy-Based Loan for Subprogram 1 Republic of Indonesia: Affordable and Sustainable Energy Transition Program.

Transport in the city could follow an "avoid-shift-improve" approach toward sustainable mobility, which implies avoiding travel in the first place. Long journeys can be reduced with effective city land use planning and behavioral interventions like car-free days. Incentives and nudges influencing individual behavior toward low-carbon options and technologies can also promote innovative low-carbon solutions across different sectors. If travel is unavoidable, it is necessary to shift to a low-carbon or more sustainable means of transport (such as nonmotorized or public transport). This can be achieved through ensuring better public transport interconnections within areas inside and outside the city. Switching to electric vehicles is a key step combined with stringent enforcement of emission verification regulations. To boost the uptake of electric vehicles, the city must develop the necessary supporting infrastructure, particularly rapid charging terminals. This may involve various forms of partnership with private sector investors. In addition, city-level financial incentives to electric vehicle owners, such as waving parking charges and municipal taxes, may be needed.

One of the critical requirements for a successful city decarbonization is to generate awareness among city inhabitants about the available low-carbon options and the national level policies and programs aimed at encouraging their uptake. More than 70% of global CO_2 emissions and two-thirds of world's energy is consumed in cities.[9] Scaling up of city-level actions could be a significant contribution to global climate goals. However, often, cities lack the technical capacity and financial and institutional support necessary to implement the low-carbon opportunities. A city-level low-carbon transition therefore requires close collaboration between city administration, provincial and national governments, and private sector investors. This collaboration is needed to develop innovative business models targeting early and mass adoption of low-carbon options, to build institutional and technical capacity for implementing low-carbon strategies, and to identify appropriate sources of finance.

9 World Bank. Topics.

6. Conclusions

This study develops a low-carbon road map for the city and its associated investment needs. With a rising standard of living, energy demand is estimated to grow strongly in the coming decades and it is important to meet this demand in as environmentally supportive a manner as possible. As a guide, this study applies the MESSAGEix model to develop a low-carbon pathway for the city energy system. Using a scenario-based approach, two scenarios are contrasted: BAU and low-carbon. BAU assumes continuation of current and historic policy and technology trends, and low-carbon applies additional policies and technologies to support the Paris Agreement climate goals.

There is a significant potential for energy efficiency interventions in the city and these are assumed to be applied in the low-carbon scenario. Demand for space cooling is expected to grow with future economic growth. Deployment of efficient air-conditioning like inverter air-conditioners and district cooling, combined with green building design, can reduce energy needs, and limit the growth of demand in buildings. Electrification of end uses is also possible in transport through greater use of electric vehicles and residential cooking through electricity-based cooking. There are multiple benefits associated with electrification, including efficient utilization of energy, no emissions at the point of usage, and the possibility of using decentralized and local renewable energy generation. However, for these to be achieved fully electricity generation, itself, must have its carbon footprint reduced.

Under the low-carbon scenario, by 2050 the TFED is projected to be 1.81 Mtoe, which is approximately 10% lower than in the BAU case. Of the total energy demand reduction (relative to the BAU scenario) most (73%) is in transport, followed by residential non-cooking (23%). In addition to the reduction in energy demand, in the low-carbon scenario, there is also fuel-switching in the energy mix arising from the decarbonization of energy supply and a modest growth in solar PV supplies.

Emissions rise under both scenarios, but with energy efficiency interventions on the demand sectors and increased penetration of clean energy in the overall energy supply they are projected to be 18% lower in 2050 in the low-carbon scenario. The climate-related benefits from this scenario are associated with higher investment costs. The low-carbon scenario requires investments worth $3.54 billion (in 2010 prices) between 2020 and 2050, which is an additional investment of $0.48 billion over the investment requirement in the BAU scenario (16% more). Finding the funding to meet this enhanced investment program will be a major challenge for the city administration. Although the city government cannot be expected to raise the funds itself, it has to identify funding opportunities and to create the investment climate to ensure that the necessary investment is forthcoming.

For investment in low-carbon interventions, one of the dominant investments—developing solar power generation capacity (utility and rooftop scale)—cannot be implemented solely by regional policy by the Denpasar City Government alone. The role of the central government is highly anticipated by regional governments regarding investment regulations and procedures that directly involve the community in residential areas.

The involvement of the private sector is also very influential in being able to provide subsidies for the initial investment carried out by the community (currently installing a solar panel power plant in Denpasar City is still very expensive). An operational expenditure model is recommended for the city wherein the central government, renewable energy service company or private investors invest in installing solar panels and the community will only be charged for operational costs.

Similarly, in the transport sector, public shift toward using electric modes of transportation should be strengthened by support from the private sector in terms of reducing the production of vehicles using fossil fuels and accelerating the provision of supporting infrastructure for the operation and maintenance of electric vehicles.

Another critical requirement for a successful city decarbonization is to generate awareness among inhabitants about the available low-carbon options and national level policies and programs aimed at increasing their uptake. The city administration needs to support implementation of national initiatives and programs, speed up its own approval system, and promote best practices. A city-level low-carbon transition therefore requires close collaboration between city administration, Provincial and federal government and private investors should be encouraged to build institutional and technical capacity for implementing low-carbon strategies, and to identify appropriate sources of finance.

www.ingramcontent.com/pod-product-compliance
Lightning Source LLC
Chambersburg PA
CBHW041800220926
51847CB00014B/270